AF270241

ENDANGERED

BIRDS
AROUND THE WORLD

BY LISA J. AMSTUTZ

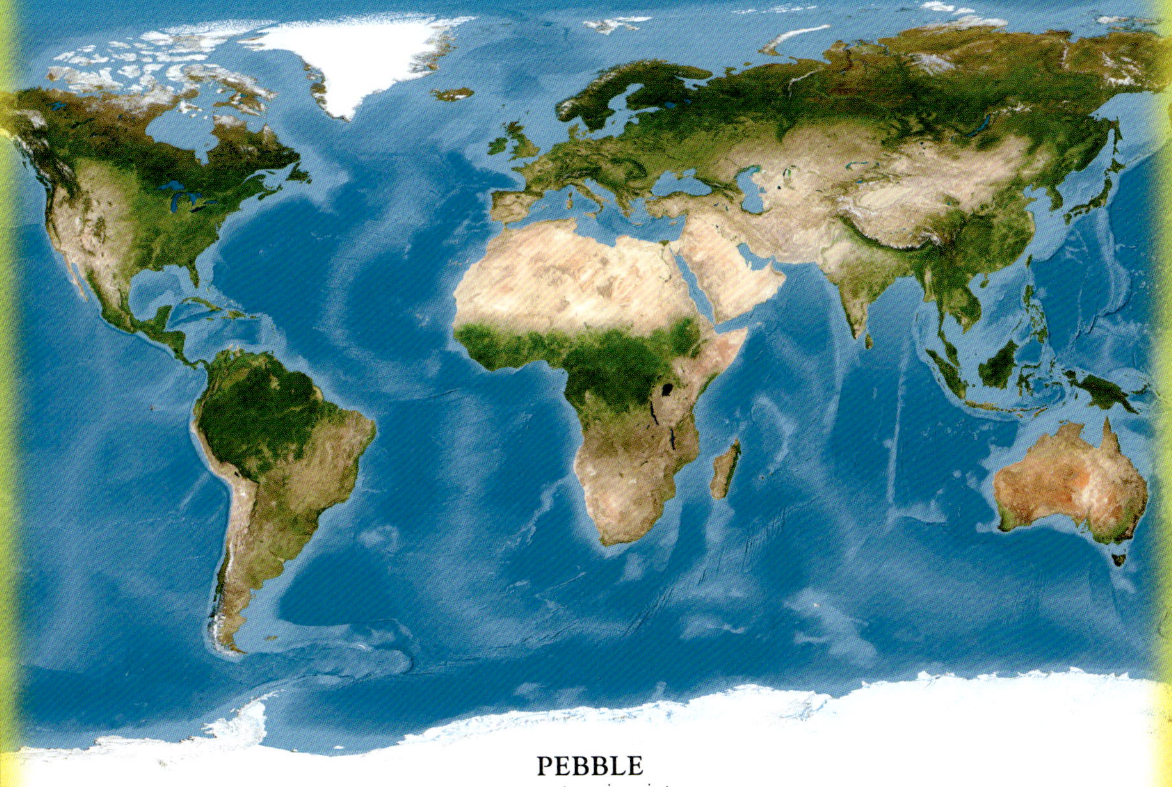

PEBBLE
a capstone imprint

Published by Pebble, an imprint of Capstone
1710 Roe Crest Drive, North Mankato, Minnesota 56003
capstonepub.com

Library of Congress Cataloging-in-Publication Data is available on the Library of Congress website.

ISBN: 9780756578329 (hardcover)
ISBN: 9780756578473 (paperback)
ISBN: 9780756578480 (ebook PDF)

Summary: From owls that live underground to penguins that build their nests in bat poop, these endangered birds are having a tough time. Learn about some incredible birds that need our help to survive.

Editorial Credits
Editor: Ericka Smith; Designer: Sarah Bennett; Media Researcher: Svetlana Zhurkin; Production Specialist: Katy LaVigne

Image Credits
Alamy: Neil Bowman, 19, North Wind Picture Archives, 9; Getty Images: Juan Carlos Vindas, 18, Kevin Schafer, 21, Robin Bush, 23, Satoru S, 26, View Stock, 27; Shutterstock: Albert Beukhof, 14, AndreAnita, 5, asantosg, 7, Bruno Martins Imagens, 15, Dennis W. Donohue, 8, Jaclyn Vernace, 13, juerginho, 6, Keneva Photography, 25, Lisa A. Ernst, 20, Piotr Krzeslak, 4, Scott E. Nelson, 11, Sergey Uryadnikov, cover, Susana Miranda, 12, Vasily Vishnevskiy, 29, Viacheslav Lopatin, 1; U.S. Fish and Wildlife Service: 16, John Magera, 17, Steve Hillebrand, 10

Printed and bound in China. 5827

TABLE OF CONTENTS

Words in **bold** are in the glossary.

All About Endangered Birds

What Is a Bird?

What do all birds have? Feathers! But not all of them can fly. Some birds, like penguins and emus, are flightless.

Birds lay eggs in nests. They warm the eggs with their bodies. Most feed their young when they hatch.

Birds can be found almost everywhere on Earth. They live in steamy jungles and icy **tundras.**

Why Are Birds Endangered?

Birds need food to eat and clean water to drink. They need places to nest. But people are destroying their **habitats**. They cut down trees. They **pollute** water. People also hunt birds for food and sport. Their pets can kill birds too.

Some birds are in danger of dying out. They are **endangered**. Without our help, they may go **extinct**.

Where Do Endangered Birds Live?

There are endangered birds all over the world. Many live in the **tropics**— areas near the equator.

Here's where you can find the birds you'll learn about in this book!

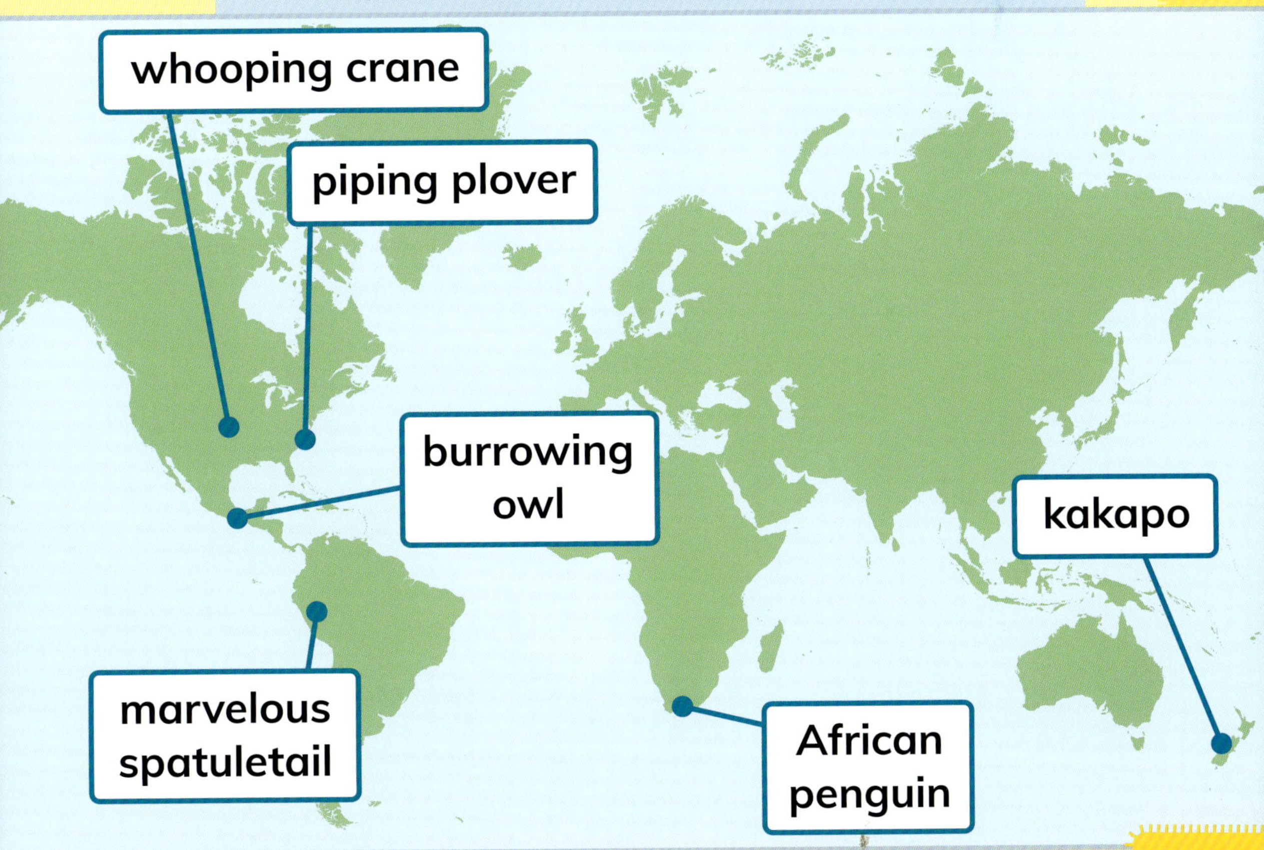

Whooping Crane

A huge bird leaps up and down. He's looking for a mate. He calls out. *Whoop! Whoop!*

The whooping crane is the tallest bird in North America. It is 5 feet (1.5 meters) tall. Its wings can be 7.5 feet (2.3 m) across.

Thousands of whooping cranes once nested on **prairies**. In the mid-1800s, settlers came to the area. They drained wetlands to create fields for planting. They also hunted cranes for their meat and feathers. By 1941, there were only about 20 cranes left.

Groups pitched in to save the cranes. New laws stopped people from hunting them. Scientists hatched eggs. They also taught young birds to migrate. The whooping cranes' numbers started to grow. Today, there are about 600 of them.

A scientist working with whooping cranes

Piping Plover

A sand-colored bird dashes down the beach. It stops and taps the sand with one foot. *Snap!* It grabs an insect.

The piping plover nests on beaches in central and eastern North America. But people like beaches too. Vehicles crush the birds' eggs. Dogs and cats eat their eggs and chicks. And homes and roads ruin their habitat.

Today, groups clean up the beaches where plovers live. They catch outdoor cats. And they put up signs to protect plover nests.

Burrowing Owl

Most owls live in trees. But not burrowing owls! They live in holes in the ground. Some dig their own holes. Some use burrows made by other animals. These owls live in parts of North and South America.

But burrowing owls are endangered in some places. Roads and towns have cut into their habitat. **Pesticides** kill them. And outdoor cats prey on their eggs and chicks.

People building a burrow for an owl

But there is hope for these owls. Some people make burrows for them from plastic pipes. They move the owls to these homes. Others work to protect the owls' habitat.

Marvelous Spatuletail

Zip! A marvelous spatuletail speeds by. It is a rare sight. This bright hummingbird lives in the forests of Peru. But it is in danger. Every day, people cut down trees for wood. They clear land for farms. This bird's home is shrinking.

A reserve for marvelous spatuletails

There are now fewer than 1,000 spatuletails left. They only live in a small part of Peru.

But some groups are trying to help. They are buying land. Spatuletails can live safely in these **reserves**.

African Penguin

A penguin waddles by. But this isn't Antarctica! It's southern Africa!

African penguins are endangered. They dig their nests in **guano**. But people take guano to help grow crops, so the birds do not have places to nest.

Nest boxes for the African penguin

Overfishing and climate change make it harder for the penguins to find food. And oil from spills can coat their feathers. Then they cannot swim.

But zoos help protect their habitat. Zoos raise penguins too. Other groups make nest boxes. They also clean up oil spills.

Kakapo

The kakapo lives in New Zealand. It is the world's heaviest parrot. This bird cannot fly. It comes out at night to climb trees and find food.

But the kakapo is dying out. People cut down trees where the kakapo lives. They also brought cats, rats, and **stoats** to the area. These animals eat kakapos.

Scientists are helping. They are moving kakapos to islands where they have removed **predators**. They are also feeding chicks to help them survive.

Making Progress

Bald Eagle

In the early 1900s, many farmers killed bald eagles. They thought the birds were snatching their chickens and lambs. They wanted to protect their livestock.

In the 1940s, a pesticide called DDT started to kill eagles too. It made their eggshells weak. The eggs broke before the eaglets could hatch.

In 1972, DDT was banned. Laws stopped people from shooting eagles. Slowly their numbers grew. In 2007, the bald eagle was taken off the endangered species list.

Crested Ibis

The crested ibis once lived in many parts of Asia. It fed on fish, snails, and crabs in **rice paddies**. Then farmers started using pesticides. The chemicals killed fish and other animals. The crested ibis had no food. By the 1960s it was nearly extinct.

In 1981, scientists found seven crested ibises in China. People worked hard to protect them. They stopped using pesticides. They set aside forest areas for the birds. Now there are more than 4,400 crested ibises there.

What You Can Do

Human activities can cause problems for birds. But you can help! Here are some things you can do:

» Don't get wild birds as pets or release pets into the wild.

» Help keep waterways clean by not littering.

» Provide food and clean water for backyard birds.

» Join a beach or river cleanup.

» Spread the word!

GLOSSARY

endangered (en-DAYN-juhrd)—at risk of dying out

extinct (ek-STINGKT)—no longer living

guano (GWAH-noh)—dried bird or bat droppings

habitat (HAB-uh-tat)—the home of a plant or animal

pesticide (PES-tuh-side)—a poisonous chemical used to kill insects, rats, or fungi that can damage plants

pollute (puh-LOOT)—to make dirty or unsafe

prairie (PRAIR-ee)—a large area of flat or rolling grassland with few or no trees

predator (PRED-uh-tur)—an animal that hunts other animals for food

reserve (ri-ZURV)—land that is protected so that animals may live there safely

rice paddy (RAHYS PAD-ee)—a wet field where rice is grown

stoat (STOHT)—a small mammal that is related to the otter and the weasel

tropics (TROP-iks)—a warm region near the equator

tundra (TUHN-druh)—a cold area where trees do not grow; the ground stays frozen most of the year

READ MORE

Amstutz, Lisa J. *Endangered Amphibians Around the World.* North Mankato, MN: Capstone, 2025.

Cheeseman, Polly. *All About Birds: An Illustrated Guide to Our Feathered Friends.* London: Arcturus, 2022.

Hickman, Pamela. *Birds.* Toronto: Kids Can Press, 2020.

INTERNET SITES

National Geographic Kids: Birds
kids.nationalgeographic.com/animals/birds

NestWatch
nestwatch.org

The Royal Society for the Protection of Birds: For Kids
rspb.org.uk/fun-and-learning/for-kids

ABOUT THE AUTHOR

Lisa J. Amstutz is the author of more than 150 children's books. A former outdoor educator, she holds degrees in biology and environmental science. Lisa enjoys learning fun facts about science and sharing them with kids. She lives on a small farm with her family.